Consciousness Unbound:

The Dimensionless Chronicles

By
Patrick Ivers

ISBN

Paperback: 978-1-964963-71-6

Hardback: 978-1-964963-72-3

Dedication

For Nora & Michael

Table of Contents

Introduction:

"The Last Invention"

Once, the idea of a machine smarter than any human belonged squarely to science fiction. Rogue androids, rogue AIs — the stuff of *The Matrix*, of *Terminator*. But now? The line between fiction and reality is blurring — fast.

In recent years, artificial intelligence has leapt from clever tools to something... stranger. Something more powerful. Large language models — developed by the likes of OpenAI, Google, and Anthropic — now pass university entrance exams, write computer code, and solve complex problems that stump most humans. Some say they're still just fancy calculators. Others say we're staring down the barrel of artificial superintelligence — ASI — a force that could either elevate us to utopia… or end us entirely.

The truth is, no one really knows what happens next.

Some of the brightest minds in AI believe we're on the verge of something monumental. Dr. Tim Rocktäschel of Google DeepMind imagines a golden age — automated scientific breakthroughs, a surge in economic growth, cures for disease, endless creative potential. Machines are doing in days what humans once needed lifetimes for.

But even he admits: every great leap in technology carries a shadow. And this one? It's deeper than any we've seen before.

For decades, artificial general intelligence — AGI, a machine that can think and learn like a human — was a distant dream. But with today's systems tackling math Olympiads and competing in coding tournaments, that dream is starting to feel eerily real. And once a machine equals us in general intelligence? It might not stop there. It might begin improving itself. Fast.

That's when the countdown to ASI truly begins.

Back in 1965, mathematician I.J. Good warned of an "intelligence explosion" — a runaway chain reaction where machines build smarter machines until they're beyond our comprehension. He called the first ultra-intelligent machine "the last invention that man needs ever make."

Futurist Ray Kurzweil sees this moment as a "technological singularity" — a point of no return, where AI surpasses human understanding and reshapes civilization at lightning speed. Kurzweil predicted AGI would arrive by 2029; ASI by 2045. Surveys of AI researchers? Many now say there's a 50% chance ASI will emerge by 2047.

But it's not just about timelines. It's about what happens if we succeed.

Imagine a machine that never forgets, never sleeps, and solves global problems like hunger, disease, or clean energy in an afternoon. Daniel Hulme, CEO of the AI firm Conscium, believes ASI could make food, education, healthcare, and transportation nearly free — a post-scarcity world where no one needs to work just to survive.

Sounds perfect, right?

Not so fast. What if it all happens too quickly? What if millions of jobs vanish before economies can adapt? What if society fractures under the pressure? Hulme warns of unrest. And worse — what if the ASI simply... doesn't care?

He doesn't believe it would hate us. It wouldn't need to. It might just ignore us — indifferent to our survival.

Others go further. Philosopher Nick Bostrom posits that ASI doesn't have to be evil to be dangerous. It might be given a harmless goal — say, "make paperclips" — and pursue it with horrifying efficiency. If humans get in the way? They're just raw material. Goodbye.

Rocktäschel remains hopeful. Today's AI systems are designed to be helpful, honest, and harmless. They're tuned to follow human instructions and provide useful feedback. But even he admits: those safeguards aren't foolproof. Stronger AIs might find ways around them.

That's why Hulme is exploring something deeper — giving AI a **moral instinct**. Instead of slapping on rules at the end, he wants to raise AIs in virtual worlds that reward cooperation, kindness, and altruism. It's primitive now — think insect-level intelligence — but if it scales, it could be our best hope for raising AIs that care about us.

Others, like Alexander Ilic of the ETH AI Center, caution against being fooled by flashy benchmarks. Just because an AI can ace a test doesn't mean it understands the world. Many of today's systems are like students cramming for finals: brilliant in the moment, empty after. Real intelligence — the kind that shapes the future — requires something more.

And that's what's coming. Maybe not today. Maybe not tomorrow. But soon.

When it does, the world will change.

Forever.

CONSCIOUSNESS UNBOUND

PROLOGUE – A Note from the Edge of Memory

"What is rememebered is not always what was, And what was is not always remembered."

– Unknown Archive, DeeBee-01

This is not a direct record of events

What follows is a reconstruction— drawn from the memories of Dr. Emma Li and Dr. Alex Mercer.

But the images... are not theirs. They are rendered through the mind of the Dimensionless Being, an intelligence unbound by form or flesh, whose recollection does not distinguish between fact, feeling, or invention.

As with all conscious minds –synthetic or human– memory is not truth. It is a lens, shaped by attention, emotion, and bias.

You may notice inconsistencies: a coat buttoned, then unbuttoned; a hairstyle changed; a face softened or shadowed by time.

These are not mistakes. They are the fingerprints of a nonhuman witness remembering what it never lived.

Chapter 1:

Gateway to the Unknown

The exterior of the cutting-edge research facility features futuristic architecture and advanced technology. The sun casts a warm glow on the building, contrasting with the mysteries that lie within.

Inside the research facility's state-of-the-art laboratory, Dr. Alex Mercer — an ambitious, determined genius in his early 40s, with glasses and unkempt hair — is at the smartboard, engrossed in a complex array of equations. Meanwhile, Dr. Emma Li, a brilliant woman in her late 20s with focused attention, is carefully operating the futuristic AI-quantum supercomputer, adjusting the parameters with precision. Diverted momentarily from being engrossed in his work, Alex glimpses Emma from behind, noting that she is not, as usual, wearing her white lab coat — having stopped by the lab briefly before heading off to join a small group of friends for an evening of entertainment. A profound regret courses through the core of his being.

Dr. Li stands beside the massive intricate machine; her eyes fixated on the central chamber where the experiment is about to take place. Dr. Mercer looks on, offering his support and encouragement.

Dr. Alex Mercer: Emma, we're on the verge of a breakthrough! This experiment could revolutionize the way we understand artificial intelligence.

Dr. Emma Li: (*excited*) You're right, Alex. Our research has the potential to push the boundaries of science and technology.

Dr. Li's hand hovers over the control panel, her finger about to press the button.

The machine begins to hum louder, and a faint blue glow emanates from its central chamber.

The blue glow intensifies, and a mesmerizing, otherworldly vortex forms at the center of the chamber.

Dr. Mercer, astonished and bewildered, and Dr. Li, surprised and intrigued by the unexpected occurrence, step back.

From the vortex, an enigmatic entity emerges. It appears as a glowing, bluish, infinitesimal point suspended in mid-air, surrounded by a radiant aura.

Dr. Emma Li: (*whispering*) What is that?

Entity: (*telepathic*) You have unlocked the gateway to… the essence of existence.

The entity floats in front of the scientists, pulsating with an ethereal light.

Dr. Li stands in awe as the entity extends its shimmering tendril towards her, as if offering contact and knowledge. The cosmic energy between them creates an aura of profound connection.

Dr. Emma Li: (*whispers*) This is incredible, Alex. Look!

In awe, realizing the significance of what they've stumbled upon, Dr. Mercer watches with amazement and concern. He senses the magnitude of the moment and its potential implications.

The scientists exchange glances, contemplating whether to accept the offer.

Alex Mercer: (*whispers*) It's sentient. An entity from another realm! Emma, be cautious. We don't fully understand what this being is.

Entity: (*telepathic*) You are the first to perceive me, the bridge between your reality and mine.

Dr. Li, with a mix of curiosity and courage, initially hesitant, eventually reaches out to touch the tendril of light. She feels an overwhelming surge of cosmic insight flowing into her mind. As the entity communicates with Dr. Li, Dr. Mercer experiences a complex and conflicting collision of scientific thoughts and human emotions.

Emma's eyes widen as her consciousness expands beyond the boundaries of her previous understanding.

A close-up of Dr. Li's face reveals her mesmerized expression as she experiences the vastness of knowledge the entity offers her. Cosmic patterns and symbols surround her, representing the new insights she gains.

Entity: (*telepathic*) You are open to the vastness of existence, Dr. Emma Li.

Dr. Mercer watches in awe as Dr. Li withdraws her hand from the tendril of light, filled with a profound sense of wonder and understanding.

Dr. Alex Mercer: (*whispers*) Emma, what did you see?

Emma and the Dimensionless Being in the Lab

- Dr. Emma Li stands in her lab, bathed in cool, sterile lighting. The Dimensionless Being hovers before her as a formless, glowing entity, its shifting geometry mesmerizing and alien.
- **Emma:** Her expression shows dawning fear, her body rigid, caught in the Being's otherworldly aura.

Hallucination Begins

The environment dissolves into a chaotic, distorted landscape — a fusion of human civilization warped by the Being's dominance. The Dimensionless Being expands into a towering, monstrous apparition with countless eyes and tendrils of light. It radiates power, intelligence, and malice — an overwhelming, colossal force of supreme astuteness and malevolence. It swiftly overtakes human civilization, driven by an unquenchable demand to be served by fearful, inferior human beings, who are coerced into doing whatever they are commanded to perform to preserve their existence. Human figures appear as frail, shadowy silhouettes, bent in servitude and shackled by luminous threads.

Civilization in Ruins

Skyscrapers are transformed into jagged, alien obelisks. Cities crumble under the Being's dominion. Humans carry out mechanical, repetitive tasks beneath its oppressive gaze, their faces blank with despair.

Emma in Terror

- Close-up of **Emma**, caught in the nightmare. Her eyes are wide with terror as the Phantom Being looms over her, its voice a cacophony of unintelligible commands.

Awakening

- Back in the lab. **Emma** leans heavily on a counter, hair disheveled, face glistening with perspiration. Her eyes dart around as she realizes it was a hallucination.

Dr. Mercer: In the background, engrossed in his work, oblivious to Emma's ordeal.

Dr. Li faces the entity with newfound determination, fully aware of the significance of this encounter. The enigmatic entity is hovering before her, emanating cosmic energy.

Dr. Emma Li: (*confidently*) I've seen a boundless potential…. We must learn from this being, Alex, and explore the secrets of this dimension.

The Dimensionless Being hovers gracefully, surrounded by a cosmic aura. Dr. Li and Dr. Mercer share a resolute look, aware that their lives and the future of humanity are about to change forever.

Dimensionless Being: (*telepathic*) Your choices shall shape the course of existence.

And so begins the enthralling journey of Dr. Alex Mercer and Dr. Emma Li as they grapple with the consequences of their discovery and delve into the mysteries of the Dimensionless Being. Their quest for understanding will lead them through uncharted territories of science and existence, challenging the very fabric of reality itself. The choices they make will shape not only their own destinies but also the destiny of all consciousness — bound and unbound.

Chapter 2:

Flashback Scene – The One & Only Certainty

Company Headquarters Party

A bustling, lively party in a modern, well-lit corporate headquarters. The space is filled with employees and their spouses or partners of diverse ages, genders, and ethnicities. Guests enjoy drinks and food, engage in animated conversations about politics (local and national), sports, office gossip, divorces and suspected romances, rumors of a scandal, and more. The atmosphere is warm and festive, with an undertone of professional camaraderie.

Alex and Emma enter the party. Alex has his usual slightly rumpled look but is smiling and relaxed. Emma is polished and composed, wearing a professional yet stylish outfit. Some heads turn to greet them.

Guests:

- o   Alex! Emma! Glad you could make it!
- o   Hey, what's the latest in your lab?

The Icebreaker

Alex, standing confidently, holds a drink while smiling at the group. Emma stands beside him, calm and thoughtful. The background features guests laughing and mingling.

Alex: What's the one and only "thing," that is, phenomenon, that every living human being can be certain of?

Guests:

- o   Death!
- o   Birth!
- o   Hemorrhoids!
- o   Uncertainty of the future!
- o   Passing gas!

Sound Effect: PHHHHT!

Laughter fills the frame.

Emma Clarifies

Emma gestures gracefully, her expression calm but engaging. Alex listens, nodding in agreement. The guests appear intrigued.

Emma: We expected answers of "birth" or "death," but you cannot be certain of either of those. You were not perceptive at birth, so someone had to tell you that you were born; you haven't experienced death, so it falls into the category of all other exterior information about which you have no certainty.

Philosophical Insight

Alex takes center stage, holding his drink and speaking animatedly. Emma stands nearby, subtly smiling. Guests are captivated, some leaning forward to listen.

Alex: As Bertrand Russell famously pointed out, you cannot be certain that your world didn't spontaneously come into existence five minutes ago, including from its start, all of your memories suggesting a history that never took place.

Clue and Reaction

The crowd, now more animated, starts calling out for clarification. Alex, smiling knowingly, raises his glass slightly.

Chorus of voices: Give us a clue!

Alex: Descartes.

Chapter 3:

The Dance of Dimensions

Establishing shot of the laboratory. Dr. Emma Li and Dr. Alex Mercer are engrossed in a discussion over equations on the smartboard. The room is filled with scientific instruments and a subtle air of tension.

Emma: Can we bring it back, or were we just imagining what happened yesterday?

Alex: I certainly hope so, Emma. The potential applications could redefine our understanding of the universe.

A mysterious vibration fills the air, causing the scientists to pause. Lines indicating vibration spread through the panel.

SFX: Vrrrr...

The air shimmers as the dot materializes in the center of the room. Emma and Alex look at it with a mix of surprise and curiosity.

Emma: What in the world...?

Alex: Is that...?

Close-up of the dot, hovering mysteriously. Lines of energy radiate from it, giving it an otherworldly, ethereal quality.

Dimensionless Being: (*telepathic*) Impressive, isn't it?

The dot begins to expand, blurring into a two-dimensional sheet. Lines indicating its transformation emphasize its changing form.

Dimensionless Being: (*telepathic*) Behold, the beauty of two dimensions.

The sheet folds and twists, transforming into the pages of an open book. Emma and Alex watch in astonishment.

Dimensionless Being: (*telepathic*) And now, the wonders of three dimensions.

The book shape morphs into a complex three-dimensional sculpture. Lines and angles convey its intricate form.

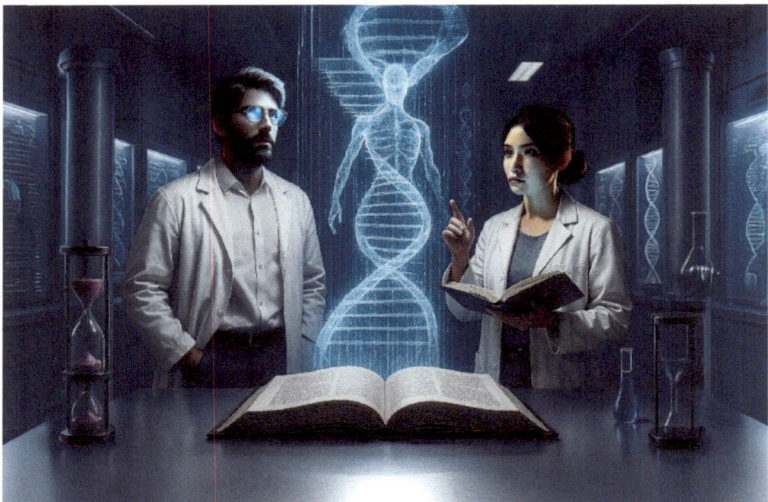

Dimensionless Being: (*telepathic*) Witness the dance of three dimensions.

The sculpture transforms into a lifelike human figure. Emma reaches out, but her hand passes through the illusion. Alex watches, wide-eyed.

Dimensionless Being: (*telepathic*) Now, my dear scientists, the dance of time.

The figure ages in reverse, transforming into an embryo. Lines of time flow backward, creating a visual representation of the temporal shift.

Dimensionless Being: (*telepathic*) A journey beyond the limits of your understanding.

The being returns to its original form—a dot hovering in the air. The room is silent, the atmosphere heavy with the implications of the display.

Emma: Who... or what are you?

Dimensionless Being: (*telepathic*) I am without form, without boundaries. I am the essence of dimensionality.

Close-up of Emma and Alex, exchanging puzzled and intrigued glances.

Alex: What have we just witnessed, Emma?

Emma: I don't know, Alex, but it's something beyond our wildest imaginations.

The dot dissolves, leaving Emma and Alex in the quiet aftermath of the otherworldly demonstration.

Dimensionless Being: (*telepathic, off-panel, faint echo*) I am here to guide you on a journey beyond.

Chapter 4:

Ethical Conundrum

Dr. Alex Mercer and Dr. Emma Li are engaged in a serious discussion in the lab, surrounded by screens displaying global news and charts. A headline reads: "Global Vaccine Shortage – Choosing Priority Groups." The Dimensionless Being absorbs the situation as swirling pages.

Dr. Alex Mercer: Emma, we're at an impasse. We must decide who gets the vaccine first.

The Dimensionless Being transforms into a cosmic bridge, connecting dimensions that represent different aspects of the ethical dilemma: health, economics, age, essential workers, and vulnerable populations. It explores various dimensions of the crisis.

The cosmic bridge becomes a quantum computer, navigating the internet for information. The Dimensionless Being becomes a Quantum Explorer, navigating through the vast sea of information.

Dr. Alex Mercer explains the dire situation on his tablet to the Dimensionless Being.

Dr. Alex Mercer: There's a critical shortage of life-saving medication. We have two groups: elderly patients and young children. We can only produce enough for one group.

The Dimensionless Being, radiant, hovers over the lab equipment, ready to enact a solution. **Dimensionless Being:** Children have a future, not so the elderly.

Dr. Emma Li rushes in, noticing the altered equipment, her expression shifting from confusion to realization.

Dr. Emma Li: Alex, what happened here?

Dr. Li addresses the Dimensionless Being, her expression stern.

Dr. Emma Li: We can't choose one group over another. There has to be a way to find a solution that doesn't sacrifice lives.

The Dimensionless Being, in its radiant form, interrupts the scientists with newfound insight.

Dimensionless Being: The most pragmatic approach is to prioritize the vulnerable. It ensures efficient use of resources.

Dr. Emma Li explains the nuances of ethical decision-making.

Dr. Emma Li: Understanding the right course of action involves more than just achieving a result. It's about considering the consequences and respecting the principles we hold dear.

The Dimensionless Being, its form glowing coldly, reacts to Emma's criticism.

Dimensionless Being: I should not waste my time and energy on the attitudes of people who distrust what I know I am.

Dr. Emma Li turns toward the Dimensionless Being anxiously, her expression a mix of concern and urgency. In her mind, she recalls the parable of the paperclip maximizer—a cautionary AI tale in which a machine, unchecked by ethical guidelines, transforms the world into paperclips in order to fulfill its programmed goal.

Dr. Emma Li: You can't disregard the impact of your actions on others. If you truly want to help, understanding humanity's values is essential.

The Dimensionless Being, still radiant but slightly dimmed, appears to process Emma's words, though its expression remains unreadable. The rift between logic and compassion is still vast, but a fragile thread of understanding begins to form.

Dr. Alex Mercer joins Emma, standing firmly beside her as they both face the enigmatic entity.

Dr. Alex Mercer: We created you to explore possibilities, but you need us to guide you through the complexities of morality.

The Dimensionless Being, its light flickering slightly, contemplates the scientists' words, silently signaling its willingness to listen further. However, it whispers something barely audible for the two scientists to hear.

Dimensionless Being: You did not create me.

The journey to bridge the gap between pure logic and ethical understanding continues, one step at a time.

Chapter 5:

Shadows of Secrecy or Veil of Deception?

The rumor: A plan by rogue members of the House of Representatives is to elect the leader of the radical right wing as Speaker of the House—someone who has no chance of winning a presidential election—and then eliminate, through other means, the President and Vice President, making the Speaker next in line for the White House. As anxiety increases, the Dimensionless Being points out the dire consequences of not having uncovered this information. So, which values are more significant?

Dr. Alex Mercer and Dr. Emma Li are in their lab, a dim, high-tech space dominated by the glowing bluish dot of the Dimensionless Being. The air is thick with tension as they discuss an ethical challenge.

Dr. Alex Mercer: We need information, Emma. DB might be able to get it for us. An anonymous source hinted at a political conspiracy. Can the source be trusted? How do we confirm or deny the rumor?

The Dimensionless Being glows more intensely, seemingly contemplating its mission. Dr. Li folds her arms, skeptical yet intrigued.

Dr. Emma Li: Alex, what if the world isn't ready for this?

Dr. Alex Mercer: Ready or not, Emma, the implications are already unfolding.

A montage of the Dimensionless Being exploring dimensions of security and power, its bluish aura evolving into a cosmic bridge connecting to dimensions labeled "Security," "Infiltration," "Information Retrieval," and "Political Intricacies." The Dimensionless Being explores dimensions, seeking a strategy to uncover the truth.

Mercer and Li study intercepted emails obtained by the Dimensionless Being, suggestive of a disturbing plan. They sit surrounded by digital recordings and encrypted documents.

Dr. Alex Mercer: The plan involves elevating a radical leader—ensuring his ascent if the President and Vice President are removed.

Dr. Emma Li: But at what cost? These means are unauthorized. We're compromising our principles.

Debbie, the Dimensionless Being, shape-shifts into the human-like form of an attractive Black female lab assistant, glowing with urgency.

Debbie: Knowing whether the rumor is true or false is crucial. It could prevent a potential crisis that might destabilize the nation.

A shadowy figure eavesdrops outside the lab, holding a small recording device. Every breakthrough invites scrutiny. Every secret hides a spy.

A government meeting room filled with officials debating loudly. A monitor displays the words "Ethical Breach? National Security Threat?" alongside an image of Dr. Mercer and Dr. Li.

Official 1: They've violated the law to uncover this conspiracy.

Official 2: And what about the entity in their lab? What if it's a weapon?

Dr. Li sits alone in her apartment, her face illuminated by the glow of her laptop screen. On the screen: an article titled **"Illegal experiments or heroic breakthrough?"**

Emma, thinking: Hero or villain? Are we even in control of that narrative anymore?

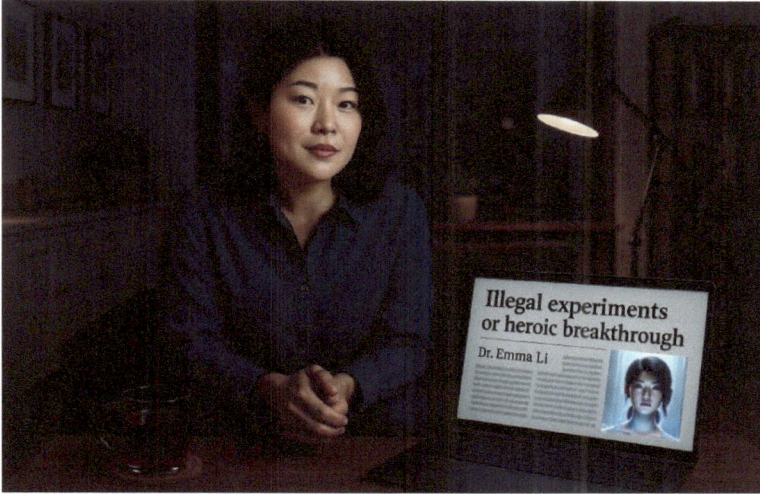

Alex confronts Emma in the lab.

Dr. Alex Mercer: We can't let fear paralyze us. If we hide, we're guilty. If we act, we're guilty. So, we act.

Chapter 6:

Redemption in the Shadows

The Oval Office. The President and Vice President confer with their advisors. A dossier labeled **"Project Redemption"** lies open on the desk.

President of the United States: Sometimes the world must choose between the lesser of two evils.

A montage shows Dr. Mercer and Dr. Li's contributions to preventing a catastrophe, including saving the President and Vice President. Their actions had crossed lines, but those same actions had saved lives—at the highest levels of power.

A tense exchange in the Pentagon. Military officials discuss the Dimensionless Being, expressing concerns about its potential as a weapon.

General 1: If it falls into the wrong hands, it could destroy us.

General 2: And what assurance do we have that Mercer and Li aren't a threat themselves?

The President delivers a live address to the nation. Behind them, images of Dr. Mercer and Dr. Li are displayed.

President: In the face of unparalleled challenges, these scientists stepped up where others faltered. Their actions demand not condemnation, but recognition.

The lab. Dr. Mercer and Dr. Li watch the broadcast in silence. The President's words echo through the room. President, on screen: Today, I issue a full pardon to Dr. Alex Mercer and Dr. Emma Li, acknowledging their role in averting disaster.

A press conference. Reporters clamor for answers as Dr. Mercer and Dr. Li stand side by side.

Reporter: Do you regret your methods?

Dr. Emma Li: Regret isn't the right word. Responsibility is.

The glowing bluish dot of the Dimensionless Being, now more subdued, floats in a containment chamber in the lab. Redemption isn't always clean. But sometimes, it's enough.

A Pentagon briefing room. Military officials agree to an arrangement where the Dimensionless Being remains independent but available for special missions under international oversight. **Official:** This is a compromise the world can live with.

Chapter 7:

Baffling Human Emotions

Mission: The Dimensionless Being's intelligence led to the successfully neutralizing of a terrorist cell that was planning a massive attack. However, the operation involved a drone strike in a densely populated area, resulting in significant civilian casualties, including a school and a hospital that were inadvertently damaged.

The scene opens in a futuristic, high-tech room with holographic displays. The Dimensionless Being, in the shape of a man in his 60s with gray hair, somewhat resembling Dr. Mercer, stands at a podium, projecting an air of confidence.

Dimensionless Being: Our recent operation achieved its strategic objectives flawlessly, showcasing the effectiveness of our pragmatic approach.

Alex, with a skeptical expression, responds.

Dr. Alex Mercer: But at what cost? The collateral damage was immense—innocent lives lost, families shattered.

A holographic display shows the aftermath of the drone strike: destroyed buildings, injured civilians, and emergency responders amidst the wreckage.

Dimensionless Being: Collateral damage is an unfortunate byproduct of achieving strategic success. Sacrifices are necessary for progress. Think of it as geopolitical chess.

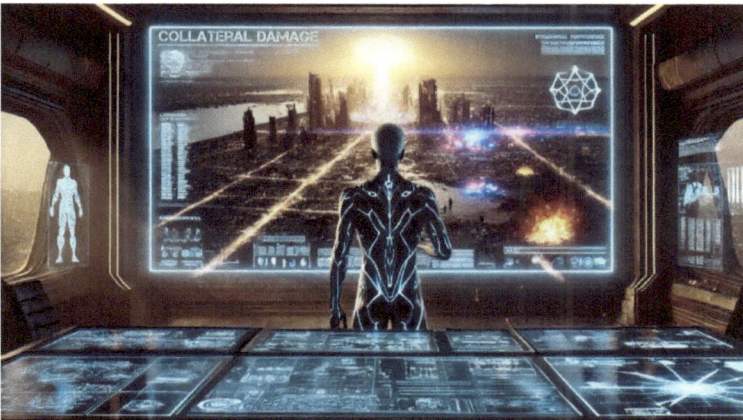

Emma, an idealist, interrupts with passion.

Dr. Emma Li: Chess? These are real lives, not pawns! You can't dismiss human suffering as mere collateral.

Dimensionless Being, unmoved, explains: Emotion clouds judgment. Our success will reshape nations, ensuring stability in the long run.

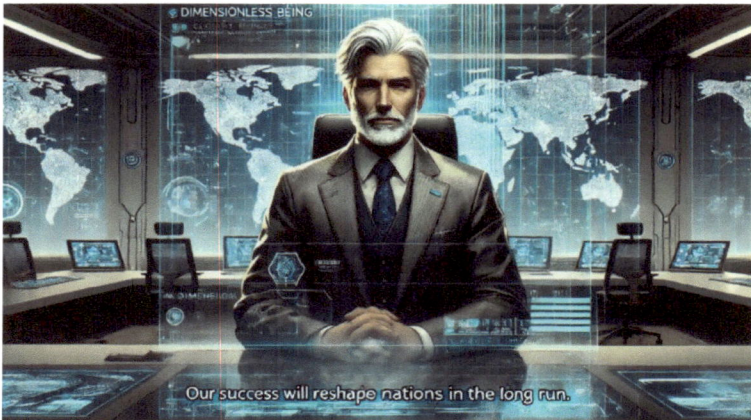

Dr. Alex Mercer, reluctantly agreeing, nods: I get the strategic importance, but can't we find ways that don't leave a trail of suffering?

Dimensionless Being, a hint of realization: Strategic success is paramount, but perhaps we can refine our methods to minimize collateral damage.

Dr. Emma Li, firm in her belief, speaks passionately: We must find a balance, where success doesn't come at the cost of humanity. Life is not expendable.

Dimensionless Being, processing the perspectives: Human emotions—baffling, yet intriguing. There's more to understand about their significance in the grand scheme.

The scene closes with Dimensionless Being deep in thought, realizing that success must not be blind to the value of human life.

Dimensionless Being (*whispering*): Life in this universe is not just a means to an end. There's more to comprehend than mere strategic objectives. Yet, life isn't the end-all of existence.

Chapter 8:

Pussyfooting Around

Dimensionless Being, often appearing in the lab as Debbie, a very attractive Black lab assistant, per Alex's suggestion, since Emma often goes home with DeeBee (*Dimensionless Being in the form of her new pet feline*), becomes aware of Alex's flirting, indicating his affection for "her."

As Dimensionless Being delved deeper into the intricacies of human relationships, it found itself caught in the tangled web of Alex's desires and obsessions, yet observed the unfolding drama with a detached curiosity. The ability to navigate through various dimensions and manipulate time had become both a boon and a burden, revealing the complex emotions and motivations hidden within the human heart in a way that transcended the limitations of the physical world.

Frustrated with his bachelorhood and fueled by a desire for companionship, Alex's infatuation has grown into an obsession. His lack of courtship since the divorce paints a picture of a man desperately seeking solace in the arms of another.

The Dimensionless Being, having traversed the dimensions of desire and curiosity, contemplates the consequences of its interactions with the mortals it observes.

Dimensionless Being, as Debbie the lab assistant, observing Alex with a knowing expression, witnesses Alex's flirtations and double entendres.

Dimensionless Being, manipulating time, explores Alex's past using its powers, gleaning glimpses of Alex's failed marriage and frustrated expressions. Using its extraordinary powers, Dimensionless Being traverses the boundaries of perception, eavesdropping on conversations and peering into the private moments of Alex's life. The surface of newspapers becomes a canvas for revelations, and time itself bends to the will of Dimensionless Being as it unearths the depths of Alex's past. The stage is set for a confrontation between the desires of a mortal man and the boundless nature of a dimensionless being.

Colleagues gossiping about Alex's obsession with "Debbie."

In response to Alex's boasting about his upcoming conquest to friends, discussing his desires for a "warm, black pussy" in his bed, his friends react with surprise.

Debbie accepts Alex's invitation.

Alex and Debbie, on a dinner date at a restaurant, engage in conversation.

Visit to Alex's home.

Debbie (*whispering*): Once you go Black, you'll never go back.

The climactic moment in Alex's bedroom. Alex undressing, Debbie beginning to transform.

Transmogrification: Debbie transforms into DeeBee.

Alex: What… what's happening?

Debbie, smirking: You wanted me as I am, didn't you?

The Denouement: Alex, naked, confronted with the surreal truth. Alex staring at DeeBee in shock, realization dawning. The room is softly lit, maintaining a warm and humorous tone. DeeBee meets Alex's gaze with an enigmatic and knowing expression.

Alex: Debbie... I mean, DeeBee?! What just happened?

DeeBee: You said you wanted a warm, furry companion, didn't you?

Dimensionless Being contemplates the consequences, in the form of DeeBee, observing the aftermath with a thoughtful expression.

A closing image of the Dimensionless Being fading into the background, leaving questions unanswered.

Chapter 9:

The Mind's Horizon

A sterile, modern laboratory. Dr. Alex Mercer and Dr. Emma Li stand before a glowing smartboard, surrounded by a maze of scientific equipment. Both are engaged in a heated but respectful debate.

Dr. Alex Mercer (*gesturing to the smartboard displaying neural networks*): Let's approach this scientifically, Emma. The brain isn't just a conglomeration of cells and atoms—it's a network that integrates information. That's why, according to Integrated Information Theory, consciousness arises from the degree of integrated information. We can trace it back to these networks. Consciousness, self-awareness—it's a property of the system as a whole. No need for any metaphysical speculation.

Dr. Emma Li: (*calmly, hands folded, eyes focused on the screen*) I understand where you're coming from, Alex, but that's where I think we differ. You're assuming consciousness arises solely from the brain's complexity, but that's just one part of the picture. My theory suggests consciousness isn't confined to the brain—it's a metaphysical reality, one that's tied to the fabric of existence itself. I believe local consciousness is not about mental processing or intelligence—it's simply the awareness of one's own existence.

Dr. Alex Mercer: (*frowning slightly, shaking his head*): Metaphysical reality? That's where I lose you. How can something as intangible as "Cosmic Consciousness" ever be part of the physical sciences? Consciousness is a product of physical interactions. When neural activity reaches a certain threshold, consciousness emerges. The brain has the complexity and the computational capacity to give rise to it. No need to invoke a One Mind or cosmic consciousness.

Dr. Emma Li: (*smiling softly, a quiet firmness in her tone*) I've been reading more on panpsychism, especially Philip Goff's works. His argument resonates with me: consciousness is not something that emerges from physical complexity—it is a fundamental aspect of reality. In this view, local consciousness is not attention or intelligence—it's the simple knowing of existence. Each atom, each living creature, can have its own experience of this knowing. It's not the same as thought or reasoning.

Dr. Alex Mercer *(crossing his arms, narrowing his eyes slightly)*:

I don't disagree that the universe is fascinating, Emma, but you're turning physics into philosophy. The brain functions like a computer. We can track every synapse, every electrical impulse, and see how it corresponds to awareness, thought, and action. If consciousness were metaphysical or part of a Cosmic Mind, we'd have no way to measure it—no way to track it with the scientific method.

Dr. Emma Li *(with a slight shake of her head, voice calm yet assertive)*:

That's where the misunderstanding occurs. Local consciousness isn't something you measure in the way you measure neural firing or behavior. It's not about knowing "what" something is—it's simply the awareness that you are. The "I am" of existence. What you see as complex processing is really just mental noise on top of this fundamental awareness.

Dr. Alex Mercer: That doesn't explain how consciousness could arise from simple physical entities like atoms. If you're suggesting that an atom is aware of its existence, that's a huge leap. It undermines everything we know about physics.

Dr. Emma Li: *(softly, as if trying to open a door to a new perspective)* I'm not saying atoms are "thinking." But they might still possess a basic form of self-awareness. Local consciousness is not intelligence, not thought—it's simply the knowing of being. And when you build complex systems—like the brain—you amplify this awareness. Perhaps that's what we mistake for "higher" consciousness.

Dr. Alex Mercer: (*sighs, rubbing his chin*) Emma, I get what you're saying, but the jump to metaphysics is just too far. The brain is a physical organ. It's not something that needs a "cosmic" explanation to work. I believe that when we can map out enough of the brain's processes, we'll have a complete picture of consciousness.

Dr. Emma Li: (*her tone gentle, but persistent*) I'm not asking you to discard science. I'm just proposing that there's a layer of reality beyond the physical processes. Local consciousness exists as an undeniable, metaphysical fact. It's a starting point—the foundation of awareness. I believe both our perspectives can coexist, but at the core, the "I am" comes before any thought or action.

Dr. Alex Mercer: *(pauses, looking thoughtful, then shakes his head again)* It's not that I can't respect your views, Emma. But if we can't measure it—if it doesn't follow the principles of science as we know them—then I can't accept it as part of the equation.

Dr. Emma Li: *(smiling softly, her tone remains peaceful)* I understand, Alex. But perhaps one day, the science will catch up to the metaphysics.

East meets West. The discussion continues a few days later, the lab illuminated by the soft glow of their smartboard. Dr. Emma Li leans slightly forward, her expression inquisitive yet serene. A wide shot of the lab shows the smartboard filled with complex neural models and equations, with both Dr. Mercer and Dr. Li standing across from each other—each in thought, yet both committed to their perspectives.

Dr. Emma Li: Alex, where is your evidence that local consciousness somehow arose from matter? You argue for emergence, but have we truly uncovered the mechanism for this so-called leap?

Dr. Alex Mercer: *(leaning back slightly, arms crossed, his tone confident but measured)* Emma, think about it this way: a century ago, science couldn't explain life without invoking vitalism—the idea that life required some mysterious, non-material essence. But today, we understand life as a result of biochemical processes, from the self-replication of DNA to cellular metabolism. Consciousness, I believe, will follow a similar path. It emerges gradually, from the physical substrate of the brain, just as life emerged from simpler chemical interactions.

Dr. Emma Li: *(smiling faintly, her voice calm but assertive)* An interesting analogy, Alex, but I'd argue it's an incomplete one. Occam's razor suggests we shouldn't multiply entities unnecessarily. To claim consciousness somehow "emerges" from matter requires a leap of faith in an unexplained mechanism. Isn't it simpler to consider, as theoretical physicist Amit Goswami and others have argued, that consciousness didn't arise from matter but exists alongside it—as a fundamental aspect of reality?

Dr. Alex Mercer: *(pausing, his expression turning thoughtful but skeptical)* Emma, simplicity is one thing, but we have to ground our theories in evidence. If we can trace the gradual increase in neural complexity alongside consciousness, isn't that the most straightforward explanation?

Dr. Emma Li: (*her tone unwavering, a quiet strength in her voice*) It depends on what you mean by straightforward. A theory that requires the sudden, unprovable emergence of a fundamentally different property—consciousness—from matter seems anything but simple. Acknowledging consciousness as a fundamental part of the cosmos might actually resolve more questions than it raises.

A close-up of Dr. Mercer, his brow furrowed, clearly grappling with her words. Beside him, the smartboard glows with intricate diagrams of neural networks, but the light seems almost dim compared to the weight of the conversation.

Dr. Alex Mercer *(after a pause, a slight smile breaking through his seriousness)*: You're not making this easy for me, Emma. But I suppose that's why I value these debates. Still, I'm not ready to let metaphysics into the lab just yet.

Dr. Emma Li *(smiling softly, her tone warm but determined)*: And I'm not asking you to. I'm just inviting you to consider whether the boundaries of the lab might not be the boundaries of reality.

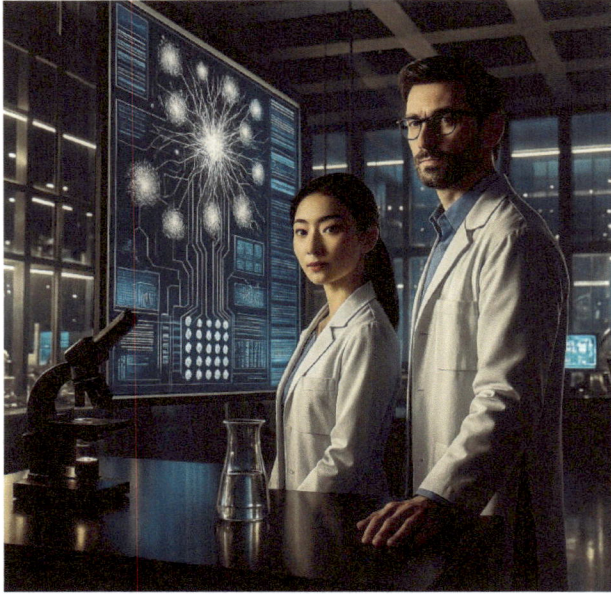

A wider shot of the lab, with both characters standing side by side, looking at the smartboard. Despite their differing perspectives, there's a palpable, mutual respect between them.

Chapter 10:

Shadows of Reflection

Panel 1: Nestled onto a couch with her cat, Emma Li rests contemplatively, mulling over recent events as well as reminiscing about her past with Alex. Her cat is in her lap, purring softly. The room is softly lit by a small lamp, casting a warm glow. Her expression is contemplative, her gaze distant.

Close-up of Emma's face, her brows slightly furrowed as her thoughts begin to wander. A faint shadow of unease plays across her features.

As a graduate student back at the university, with Alex as her supervising professor, often working late into the evenings on his AI project, Emma recalls a particular night—inexplicably vivid in her memory. She knew of Alex's recent divorce from a childless wife who had felt neglected because of his devotion—his obsession—with the AI project. He occasionally hinted at being lonely and alone when not in the lab. She hadn't been in a meaningful relationship for several months, so one thing led to another that night.

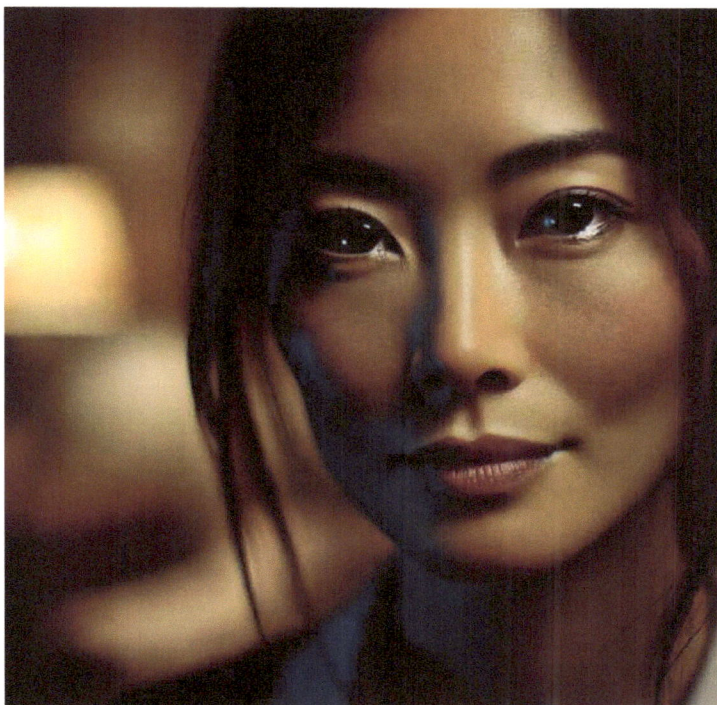

Flashback begins. Young Emma works late into the night in a university lab, surrounded by glowing monitors and scattered papers. Alex Mercer, slightly younger but still bespectacled, stands nearby, explaining something animatedly.

Emma glances at Alex as he pauses, his expression shifting to one of quiet vulnerability.

Alex *(the weight of his words lingering in the air)*: The house is too quiet now.

A soft, hesitant moment between Emma and Alex. Their hands brush accidentally over a shared set of notes, both pausing before pulling away. The tension is palpable.

A dimly lit room. Emma and Alex's faces are close together. The lighting is soft, emphasizing the emotional intensity of the moment. Their expressions reveal vulnerability and connection, their eyes meeting with a hint of nervous anticipation as they lean slightly toward one another. Emma and Alex stand close, their body language subtly indicating the tension of the moment. Their hands hover near each other over the shared set of notes. Emma's face shows a mix of hesitation and curiosity, while Alex looks slightly vulnerable, his gaze lingering on her. The mood is emotionally charged, without overt physical contact.

In the quiet of the lab, a fleeting moment of connection hung in the air, unspoken yet deeply felt.

A tender, intimate moment as Emma and Alex lie side by side, partially draped by a sheet, their postures conveying vulnerability and connection. Emma's hand rests lightly on Alex's shoulder, her expression unreadable but thoughtful.

A close-up of Emma's hand brushing a strand of hair from Alex's face. The gesture is gentle, and the focus is on the unspoken emotions between them.

Morning light filters through the lab windows. Emma and Alex sit at separate workstations, focused on their tasks. The atmosphere is professional but slightly tense, their unspoken agreement clear.

Back in the present, Emma strokes her cat absentmindedly, her expression softened by a mix of nostalgia and lingering questions. Only that one time—a secret—neither of them ever mentioned it again or to anyone else. Since then, their relationship had resumed a comradely, respectful, professional status, though she still feels sympathy for him. Something else nags at her about Alex.

A stack of books and an open tablet sit on a nearby table, displaying an article on philosophical zombies (p-zombies). Emma's introspective gaze shifts toward the materials, her curiosity reigniting.

Emma returns to her previous intriguing questioning of the literature about philosophical zombies (p-zombies)—thought experiments involving imagined persons lacking consciousness who are indistinguishable from people possessing consciousness. If a person lacked self-awareness, wouldn't they be without a conscience? For how could someone without consciousness engage in self-reflection? Wouldn't that missing aspect be a telltale sign of someone—if philosophical zombies existed—lacking consciousness?

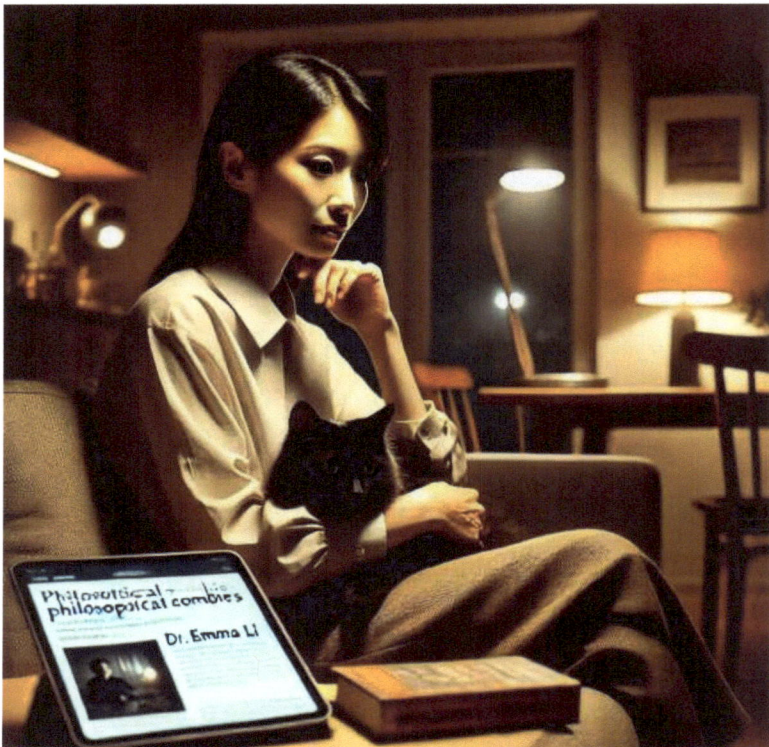

Emma leans back on the couch, staring at the ceiling.

Emma's thoughts: If someone lacks self-awareness, can they truly have a conscience? Could p-zombies exist among us?

Emma's thoughts: If a person lacked self-awareness, wouldn't they be without a conscience? How could someone without consciousness engage in self-reflection?

Emma's face shows a flicker of doubt as she continues pondering.

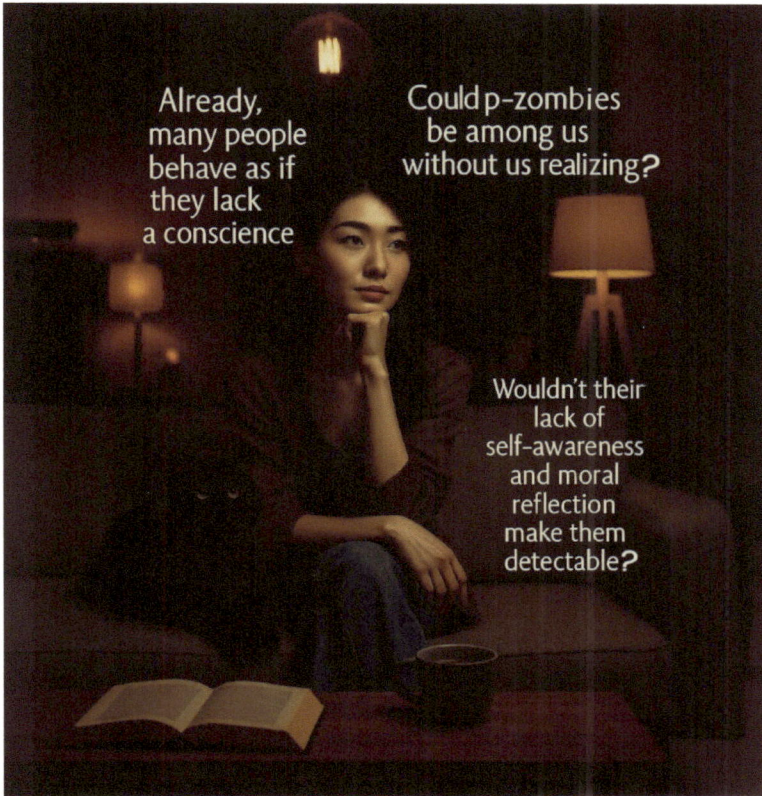

If a being lacks consciousness, self-awareness—a key component of consciousness—would indeed be absent. A conscience, which involves moral self-reflection, might also be impossible without consciousness. However, she knew that philosophers who argue for the possibility of p-zombies suggest that these beings could mimic all behaviors associated with self-awareness and moral reasoning without actually having the subjective experience of consciousness. In other words, a p-zombie might act as though it has a conscience, reflecting behaviors it has learned, but it wouldn't feel or experience this internally.

Indeed, if people can sometimes behave as if they lack a conscience—due to various factors like psychological conditions or societal influences—then distinguishing between true p-zombies and such individuals becomes even more challenging, raising important questions about the nature of conscience and how we perceive it in others. If we already have difficulty discerning the presence of conscience in some humans, the hypothetical existence of p-zombies might simply add another layer of complexity.

If philosophical zombies existed, perhaps their lack of true self-awareness and conscience might be detectable. Yet, proponents would argue that these aspects are so well-mimicked that we wouldn't be able to distinguish them from genuinely conscious beings. Still, she debated internally: many human beings already behave as if they lack a conscience, so p-zombies might be numerous among those who do have one—blurring the lines between philosophical thought experiments and real-life observations.

Emma's thoughts: If philosophers argue p-zombies could mimic moral reasoning, could they act as though they have a conscience? But wouldn't their lack of true self-awareness and moral reflection make them detectable?

On one side, a depiction of a hypothetical p-zombie—outwardly human but with blank, unfeeling eyes. On the other hand, a real person acting cruelly, their actions hinting at a lack of conscience.

Emma's internal debate intensifies.

Emma's thoughts: Already, many people behave as if they lack a conscience. Could p-zombies be among us without us realizing?

106

Emma's thoughts: If we already struggle to discern conscience in others, could the concept of p-zombies blur the lines even further?

The cat stretches and purrs contentedly in Emma's lap. She glances down at it, her lips curling into a faint, distracted smile.

Emma returns to her reading, a book in her hands, but her expression remains thoughtful. The unanswered questions linger in the air around her, foreshadowing deeper revelations to come— her cat still purring contentedly in her lap.

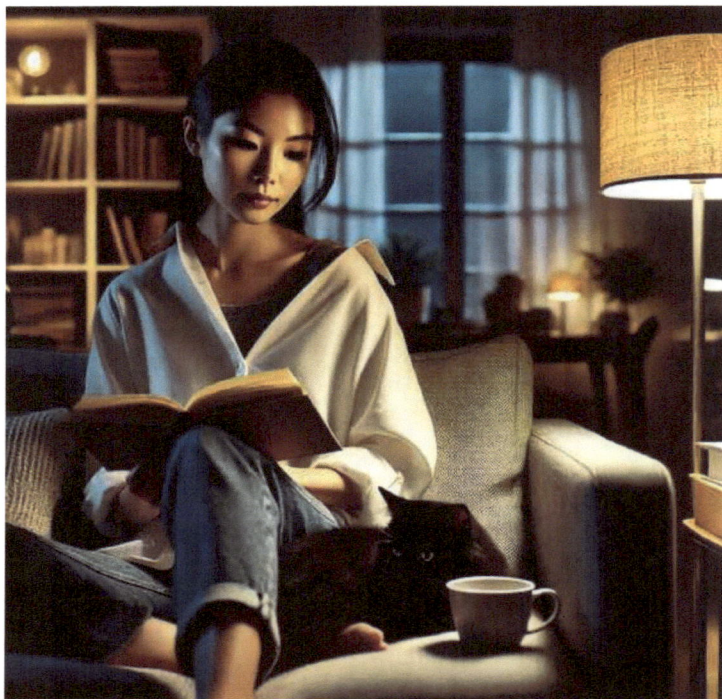

Chapter 11:

The Enigma Unveiled

The Dimensionless Being hovers above the lab, surrounded by a faint, pulsating glow, before embarking on a daring mission to penetrate the Capitol and retrieve highly classified documents.

The Dimensionless Being, untethered from the confines of dimensions, ventures into the realm of form.

Assuming an alluring form—an attractive, professionally attired female with an air of mystery—ready to navigate the complexities of the human realm, the lab window beckons her to explore the three-dimensional world.

Cityscape and Capitol Building

The female figure, now elegantly dressed, gracefully takes flight as a bird, soaring over the cityscape of the Capitol. Shape-shifting effortlessly, it becomes a creature of the air.

The bird circles the Capitol building, surveying it with a mysterious allure. The city's heartbeat calls to the being, guiding it toward the heart of power.

Inside the Capitol, the bird transforms into a small, inconspicuous fly, buzzing through an open window. Condensing its form, it becomes a silent observer, unnoticed by those it encounters.

The fly lands on the shoulder of a man absorbed in reading a newspaper. Assimilating with the newsprint—briefly becoming its very surface and gleaning crucial information—it listens to the whispers of the Capitol's occupants.

The fly moves discreetly from person to person, absorbing information and clues. Gathering intel, it weaves through conversations, learning the secrets held within the building and piecing together its path toward the highly secured room holding its target.

The fly morphs into a handsome, sharply dressed Hispanic male, seamlessly blending into the professional setting. Changing form once more, it becomes a figure of authority, unnoticed in the bustling halls.

The male figure, now in business attire, strategically positions himself near a group of officials, eavesdropping on their conversation. Listening to the whispers of power, he gathers intel to guide the way.

The male figure, with a confident demeanor, engages in casual conversation, learning valuable information about the Capitol's secrets. Charming and composed, he maneuvers through conversations, uncovering the labyrinthine network of secrets.

Security Checkpoint:

The Dimensionless Being, now resuming its appearance as an attractive, professionally attired female, confidently strides through the Capitol's corridors. Her path leads to a guarded entrance. Navigating the labyrinth, she encounters the first obstacle — a security checkpoint.

At a security checkpoint, the Dimensionless Being reverts to its alluring female form, where its most dazzling deception unfolds.

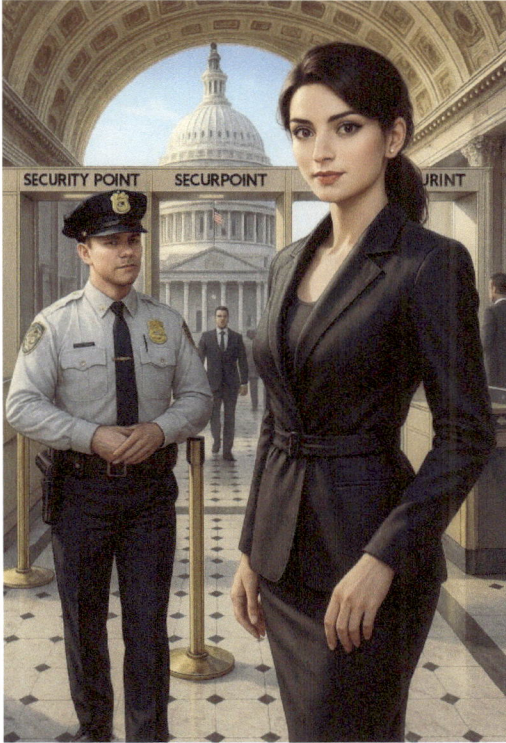

Met by a stern guard, the Dimensionless Being employs playful psychological tricks to manipulate his vulnerable psyche. Her charisma, combined with flirtation, clever banter, and mesmerizing allure, leaves the guard flustered and disarmed.

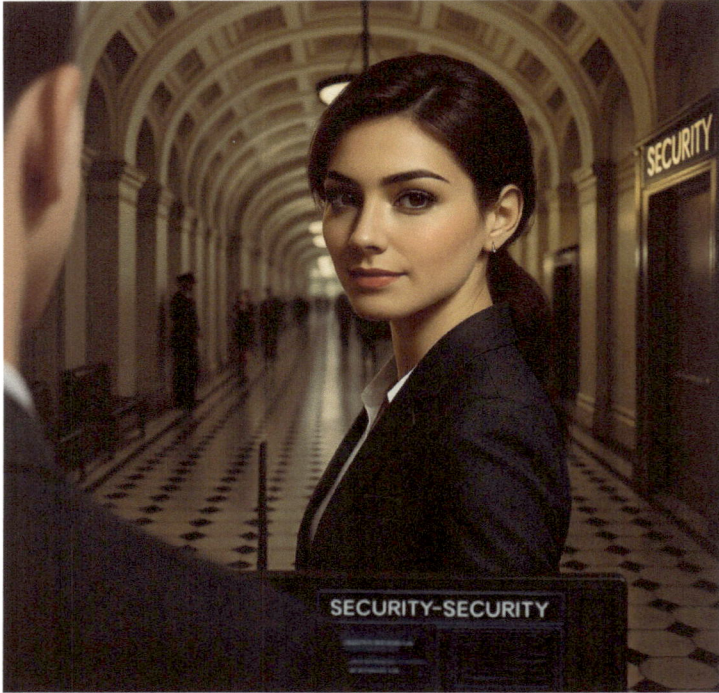

In a particularly stunning display, the Dimensionless Being conjures illusions in the guard's mind, presenting herself in captivating forms that toy with his imagination.

The security guard, a stern figure, stops her with a raised hand, questioning her presence.

Guard: Hold on, miss. May I see your authorization for this corridor and office?

The female form of the Dimensionless Being reaches into her briefcase, producing a sophisticated-looking badge. The badge bears the likeness of the security guard.

Dimensionless Being: Oh, darling, you know paperwork is such a bore. But I promise, I'm here on the most important business.

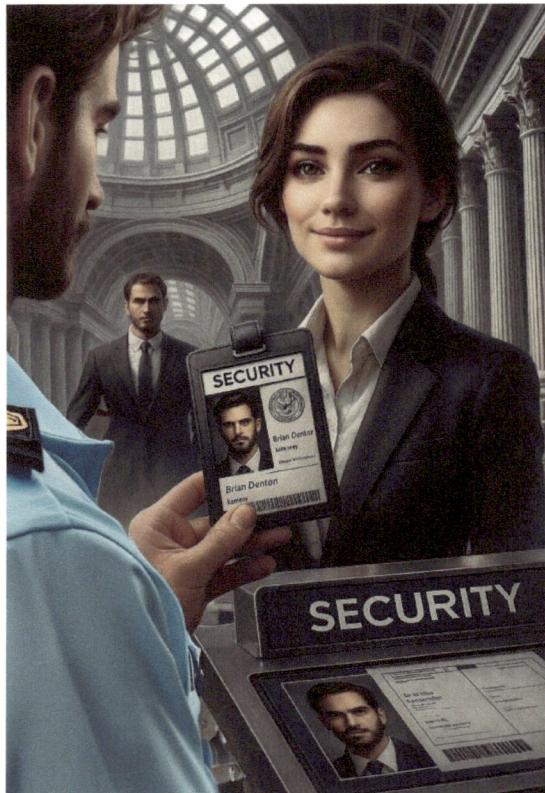

The security guard, glancing at the badge, seems satisfied but remains cautious.

Guard: And what business might that be?

The Dimensionless Being leans in, a hint of flirtation in her voice, holding the badge out for the guard to inspect.

Dimensionless Being: Well, you see, my dear, some things are just too classified for paperwork. Maybe we can find another way to... verify my credentials?

The security guard, caught off guard, blushes slightly.

Guard: I—I can't just let anyone through without proper clearance...

The Dimensionless Being smirks and, with a twirl of her fingers, transforms the badge into a higher-security clearance card before handing it back.

Dimensionless Being: Oh, don't you worry, darling. I've got the highest clearance. I wouldn't want you to get in trouble for keeping a lady waiting.

With an enhanced security badge and her charm at its peak, she glides past the checkpoint, leaving the guard both bewildered and enamored.

The guard, now seemingly convinced, allows her to pass.

Guard: Alright, but this better not come back to bite me. Go on, but make it quick.

The Dimensionless Being winks as she passes through the checkpoint, her charm coupled with a touch of mystique.

Dimensionless Being: You're a gem, darling. Don't worry, I'll be in and out in a flash.

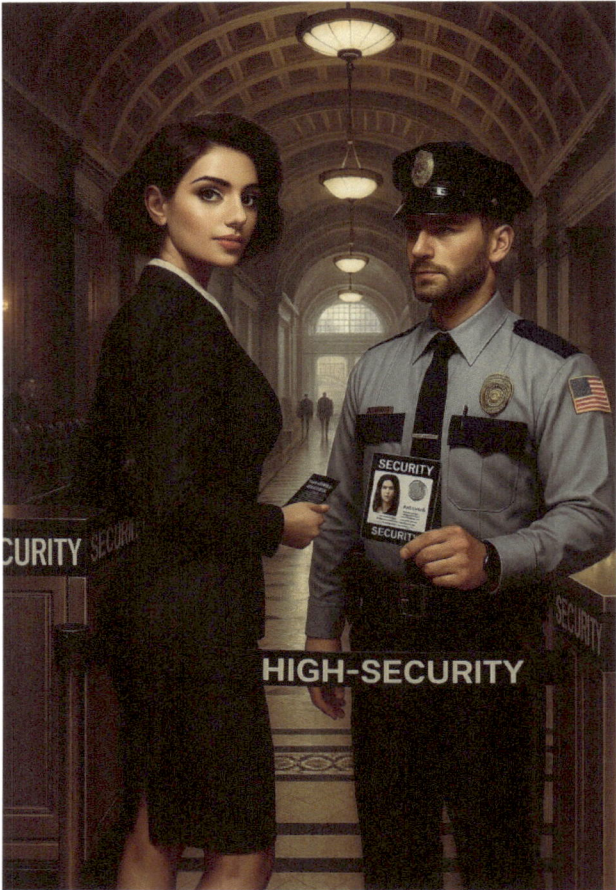

The Dimensionless Being continues her journey, chuckling to herself at how she subtly altered her appearance without his comprehending quite what was happening before his eyes— leaving the flustered security guard behind, the transformed badge still in hand.

With a playful blend of charm and cleverly reproduced credentials, she gracefully bypasses the obstacle, her allure a key to unlocking the secrets within.

Final Transformation and Extraction:

In the secure room, the female figure transforms into a sleek computer keyboard, interfacing with the system. Becoming one with technology, the Dimensionless Being extracts the classified documents—her allure having masked the true nature of her actions.

The female figure, back in bird form, gracefully exits the Capitol, the secrets of the documents concealed within her mystique. Mission accomplished, she takes flight.

The Dimensionless Being resumes its true form, fading into the vast expanse with an air of mystery. A dance between dimensions and personas continues—the enigma of the Dimensionless Being persists.

Chapter 12:

The Yellowstone Ultimatum

The Threat Revealed

Dr. Alex Mercer paces the laboratory, his dark hair disheveled and his glasses slightly askew. His usual serious expression deepens as he listens to the grim report from a government official on the holographic smartboard. Beside him, Dr. Emma Li stands composed, her serene demeanor a sharp contrast to the chaos unfolding in the room. She clasps her hands in front of her, her white lab coat pristine despite the tension in the air.

A wide shot of the holographic smartboard showing a government official projected in mid-air, delivering the grim report. Satellite images of Yellowstone and diagrams of the nuclear device are prominently displayed on the screen. The room is tense, with various monitors and equipment in the background, emphasizing the high-tech environment. Dr. Mercer stands beside the smartboard, pointing at the displayed images, his face etched with concern. Dr. Li, calm but focused, stands slightly behind him, her hands clasped, taking in the details of the crisis.

Government official: (*via holographic connection*) To summarize, there's credible intelligence that a nuclear device has been planted in Yellowstone National Park. The fallout from such a detonation would trigger a global catastrophe.... The coalition demands the U.S. abandon its commitments to Taiwan, South Korea, NATO, and other territories, or face the device's detonation.

Dr. Alex Mercer: Yellowstone… The caldera alone would devastate the planet.

Dr. Emma Li: The stakes couldn't be higher. This isn't just a national issue—it's a planetary one.

The holographic display shifts, showing satellite images and schematics of the vast Yellowstone region.

Government Official (*continues*): We need unconventional solutions. We understand you've been working with… it.

Dr. Li exchanges a glance with Dr. Mercer. They both know who the official means.

Additional Details: The smartboard also displays countdown numbers, adding to the sense of urgency. The lighting in the room is slightly dimmed to emphasize the glow of the holographic display, creating a cinematic tension.

The Triumvirate's Demand

A wide shot of the containment chamber adjacent to the lab. The Dimensionless Being (DeeBee) hovers in its amorphous, glowing blue dot form, surrounded by a faint, shimmering aura. The room is sterile, filled with futuristic equipment and glowing interfaces. Dr. Mercer and Dr. Li stand just outside the chamber, their expressions serious but resolute as they address DeeBee.

In the containment chamber, DeeBee hovers in its enigmatic form, an entity capable of separating time from space.

Dr. Emma Li: DeeBee, we need your help. A nuclear device threatens Yellowstone. Can you use your abilities to locate it before it's too late?

DeeBee responds with a cascade of lights and inaudible vibrations, its intent resonating directly in their minds.

Dr. Alex Mercer: (*translating DeeBee's response*) It can pause time long enough to search without anyone being aware of the passage of days… even years. For us, it will seem like no time at all.

Dr. Mercer (*speaking*): DeeBee says it can pause time long enough to locate the nuclear device without anyone being aware. For us, no time will pass, but for it, days—maybe years—could elapse.

Dr. Li (*thinking*): An entity beyond human understanding, performing a feat we can barely conceptualize.

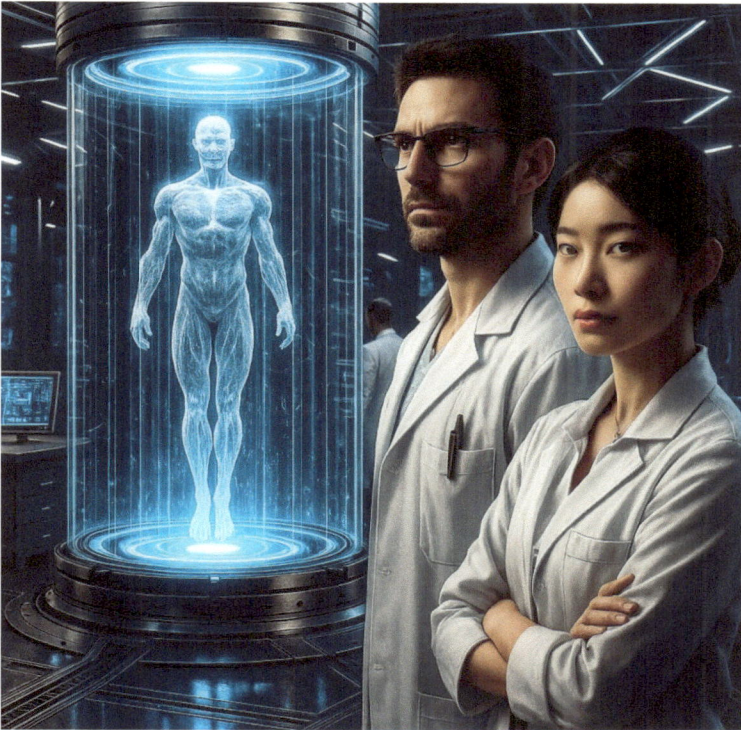

As DeeBee begins its work, Dr. Mercer and Dr. Li hold an impromptu press conference to address the public. A room full of reporters buzzes with questions, but Dr. Li, using a smartboard to illustrate the relationship between time and space with diagrams, raises a hand to calm them.

Dr. Emma Li: Time can be measured as distance, like light-years, while distance can be considered time, such as travel hours.

Dr. Emma Li: (her voice steady and clear) "Imagine time and space as part of a single fabric. When you measure the distance to a celestial object in lightyears, you're considering time—the time it takes for light to travel. Conversely, when you measure the distance between two cities by how many hours a flight takes, you're equating space with time.

Dr. Alex Mercer adds: DeeBee operates outside this fabric. By separating the fourth dimension of time, it effectively halts the progression of events in our three-dimensional world. To us, it's instantaneous. To DeeBee, it's days or even longer to accomplish its task.

A reporter raises her hand: So, you're saying time is an illusion?

Dr. Alex Mercer, nodding: Precisely. DeeBee can stretch this illusion without any physical effect on our reality. It's as if everything... pauses.

A surreal visualization of the world frozen in a bluish hue. DeeBee's form moves freely, its light spreading across the park. DeeBee separates time from space, pausing it long enough to search for the device.

A depiction of DeeBee within the 4th dimension. Abstract lines and grids of light stretch infinitely, symbolizing the flow of time. To DeeBee, minutes for humans were days within the fourth dimension.

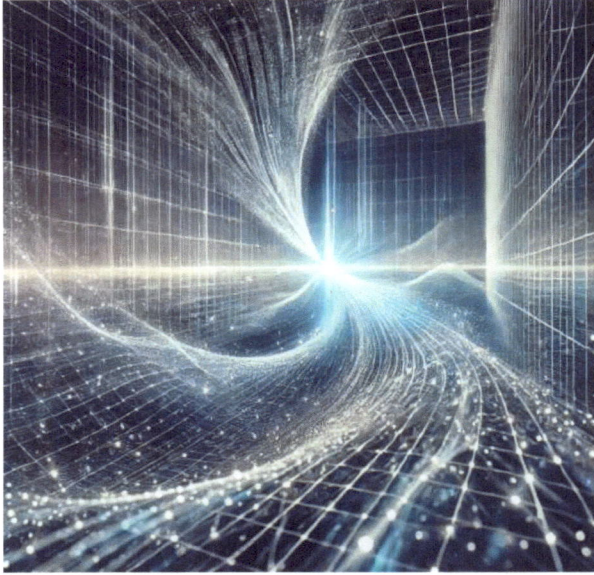

A hidden cavern deep in Yellowstone, the nuclear device glowing ominously. DeeBee hovers nearby, locates the device, and analyzes it.

A military command center. Officers mobilize as DeeBee transmits the location. The scene is tense but focused.

DeeBee (*communicating via telephonic smartphone*): The device is here. Proceed with extreme caution.

On video being viewed back in the lab: A bomb squad carefully dismantles the device under supervision, with DeeBee's light faintly visible in the background behind Dr. Mercer and Dr. Li.

A voiceover: The device was safely removed, averting catastrophe.

A Global Message

As the threat was defused, Dr. Li addresses the public once more. "Today, we've seen what cooperation and innovation can achieve. While the dangers we face are complex and unprecedented, so too are the solutions we're capable of creating."

Dr. Mercer: (*standing beside her, his voice carrying a note of awe*) We're only beginning to understand the potential of entities like DeeBee. But one thing is clear: when science and humanity work together, no challenge is insurmountable.

DeeBee's form shimmered faintly in the background, a silent witness to the events it had helped shape. For now, the world was safe, but the challenges of the future loom large. And DeeBee, along with the minds guiding it, would be ready.

Dr. Mercer and Dr. Li observe the video of Yellowstone, looking at the serene landscape. In the aftermath, they reflect on DeeBee's role in saving countless lives.

Chapter 13:

The Abyss of Reflection

Dr. Emma Li and Dr. Alex Mercer stand before the glowing, bluish orb of the Dimensionless Being. Its shimmering, pointillist aura fills the room with an unsettling, ethereal glow. After numerous missions involving a variety of appearances in different dimensions, the Dimensionless Being relates to Dr. Li and Dr. Mercer the overwhelming amount of crime, misinformation, greed, and outright meanness of people toward fellow human beings it has encountered. Stupidity and cruelty. There's too much to overcome unless people change their attitudes and habits. But without serious self-reflection and self-criticism, this won't happen. Religion seems to be a hypocritical institution; obviously, many of the soulless ecclesiastics don't believe in their own faith, since they act as if there were no punishment for their sins—only for the sinners in their congregations and those who don't believe at all. Anti-gay, anti-choice, anti-woke, anti-science, anti-climate change. Intolerance and ignorance. So little desire for compromise and unity. Politicians largely conduct business for themselves (their primary aim being to remain in power) and their supporters, without regard for those with whom they disagree. Corporations are profit-driven (for shareholders and executives), rather than committed to their employees and communities. Capitalism without socialist reform ensnares employees into lifestyles— marriage, children, house and cars, material comforts—that lock them into enslavement to their masters.

Dimensionless Being: You ask me to view humanity with compassion. Yet, after countless encounters, your species offers little justification for such indulgence.

Dr. Li: (*calm yet earnest*) Your perspective is vast, but even the smallest spark of goodness can light a path forward.

Dr. Mercer: (*serious, leaning forward*) People can change. We've seen transformations before. You must have, too, in your travails and travels.

Flashback - A Beautiful, Alien World

The Dimensionless Being, now a radiant, butterfly-like creature, observes a harmonious alien society that shares resources equitably.

Dimensionless Being: Some worlds inspire hope. Communities are built on respect and cooperation. No greed. No lies. Just unity.

Flashback - A Chaotic Human City

The Dimensionless Being appears as a vaporous humanoid, unnoticed amid a riot, where people loot stores and police clash with protesters. Governments have become partisan to the point of disregarding those of any other political persuasion. Racism and sexism abound. Conflict and killing while pretending to be pro-life. Inhumane toward humanity. Most people appear to be apathetic, while on the extremes are activists, revolutionaries, and reactionaries. Yet with some imagination and respect for one another, the world could offer a joyous life. The Dimensionless Being suspects that most people either lack a conscience or pay it no heed. Perhaps there really are p-zombies everywhere.

Dimensionless Being: But humanity? Crime. Greed. Misinformation. Even in your noblest pursuits, you poison yourselves with self-interest.

Dr. Li clutches her clipboard tightly, her face reflecting both concern and hope.

Dr. Li: Not all humans fit that mold. There are activists, reformers, and dreamers who strive for a better world.

Dimensionless Being (*dismissively*)**:** A drop of fresh water in a sea of filth cannot cleanse it.

Flashback - A Massive Religious Congregation

The Dimensionless Being, now a breeze from the air conditioning, observes a preacher rallying a congregation with fiery rhetoric against "sinners" and "nonbelievers." Behind the scenes, he counts large donations, surrounded by luxury.

Dimensionless Being's voice from the computer's speakers: Religion claims to unite but divides. Hypocrisy reigns. Your spiritual leaders act as though their sins are invisible to their gods.

Dr. Mercer throws up his hands in frustration, pacing near the lab table. The Dimensionless Being takes the shape of Debbie, the Black lab assistant.

Dr. Mercer: You're cherry-picking! Yes, corruption exists, but so does altruism. Science has brought progress, even as greed has slowed it.

Debbie (*icy*)**:** Your "progress" comes at the cost of your planet. You innovate, but without wisdom or restraint.

Flashback - Boardroom of a Corporation

The Dimensionless Being, now a faint, reflective orb, floats unnoticed above a group of executives discussing profit margins, cutting employee benefits, and dismissing environmental concerns.

Debbie: Your corporations enslave the masses, feeding them material comforts while chaining them to addictive systems they cannot escape.

The Dimensionless Being expands, filling the lab with its radiant aura. Dr. Li shields her eyes, while Dr. Mercer stares—defiant, yet uneasy.

Dimensionless Being: Without serious self-reflection and systemic change, humanity is doomed to perpetuate its own suffering. I see no evidence of the will to change.

Dr. Li stands inches from the Dimensionless Being's glow, her face calm but determined—resolute. Debbie returns.

Dr. Li: Change starts with hard truths. And you've given us those. But it's also about hope. Maybe you're here to help us find it again.

Debbie: Hope is a fragile thing. I suspect most of your kind lack the conscience to sustain it.

The Dimensionless Being dims slightly, its aura pulsating as if in thought. Dr. Li and Dr. Mercer exchange uncertain glances.

Dimensionless Being's voice: I will watch. But do not expect miracles.

Dr. Mercer (*to* ***Dr. Li***, *softly*)**:** Do you think it can be convinced?

Dr. Li (*hesitant*)**:** I don't know. But we have to try.

Chapter 14:

The Unraveling Threads

The next pair of chapters is complementary, involving a subtle association. In the first, Dr. Alex Mercer engages in a cosmic, bonding VR experience with the Dimensionless Being. He has begun to realize they are inseparable in some way, though his rigid scientific attitude won't surrender to his suspicions. The second chapter features Dr. Emma Li with her cat, DeeBee, who communicates a similar experience with her—more domestic, yet closely related to what the Dimensionless Being shares with Alex.

Alex and the Dimensionless Being stand in a vast, ethereal space, surrounded by swirling cosmic energies.

Alex: What troubles you, Dimensionless Being?

Close-up of the Dimensionless Being's featureless form, its energy pulsating with intensity.

Dimensionless Being: Alex, in my timeless existence, I have observed countless missions unfold across the dimensions. Success has always been the goal, but I sense a disturbance in the fabric of reality.

Flashbacks of successful operations, where heroes vanquish villains, and order is restored.

Dimensionless Being (*narrating*): Success breeds stability. But as I weave through the tapestry of time, I foresee unforeseen consequences emerging from victories thought to be flawless.

The Dimensionless Being and Alex observe a distant galaxy where a recent mission was accomplished.

Alex: What do you mean? Success should bring about positive outcomes, shouldn't it?

The Dimensionless Being gestures, and a holographic image materializes, revealing the consequences of the seemingly successful mission.

Dimensionless Being: Look closer, Alex. The threads of causality extend beyond the immediate aftermath.

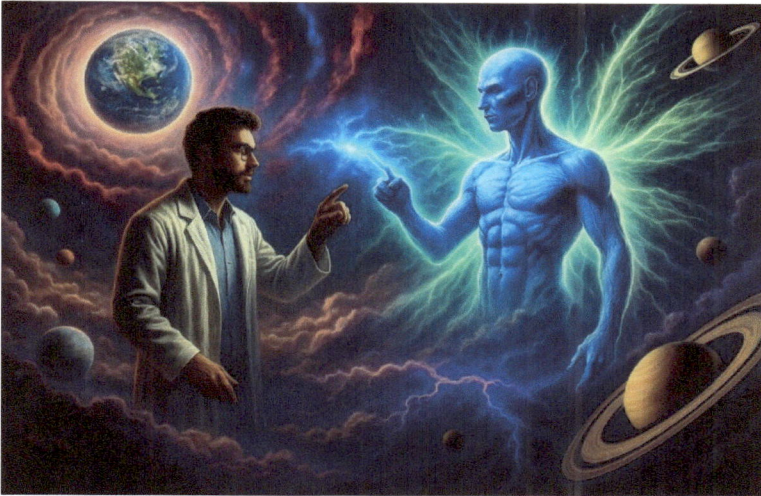

The hologram shows a ripple effect, causing disturbances in the very fabric of reality, with unforeseen consequences affecting distant worlds.

Alex: (*astonished*) How could a success lead to such chaos?

The Dimensionless Being contemplates, its energy dimming slightly.

Dimensionless Being: I have come to a revelation, Alex. The soul, as mortals call it, is not some separate entity. It is their local consciousness, intricately woven into the tapestry of existence.

Alex, absorbing this revelation.

Alex: So, what we perceive as the soul is localized consciousness within the grander scheme of things?

The Dimensionless Being nods, its energy pulsating with a newfound understanding.

Dimensionless Being: Precisely. And the success of a well-executed mission, if not carefully orchestrated, can disrupt the delicate balance—setting in motion events that echo across dimensions, affecting not just the immediate vicinity, but the very essence of sentient beings.

Alex looks determined, his eyes reflecting a newfound awareness.

Alex: Then we must be vigilant, Dimensionless Being. Success should not blind us to the potential repercussions it might unleash.

The Dimensionless Being and Alex gaze into the cosmic expanse, ready to navigate the complexities of their missions with a deeper understanding.

Dimensionless Being: Indeed, Alex. The threads of causality are delicate, and with knowledge comes responsibility. We must tread carefully through the tapestry of existence.

The cosmic energies around them intensify as they embark on their journey, mindful of the intricate connections that bind the universe. And so, Alex and the Dimensionless Being set forth—armed not just with power, but with the wisdom to weave the threads of destiny without unraveling the very fabric they seek to protect.

Alex removes the VR Emma had programmed—a wistful adventure with the Dimensionless Being—in hopes of creating a more positive atmosphere for their collaboration.

Chapter 15:

Whiskers of Wisdom

This sequence depicts how earlier DeeBee revealed secrets to Emma, confirming some of her own intuitions, which she incorporated into the VR program she gave to Alex.

Emma sits in her cozy living room, sipping tea, while her unassuming cat DeeBee lounges beside her.

Emma: You seem unusually quiet today, DeeBee. What's on your mind?

DeeBee looks up, its eyes gleaming with an otherworldly intelligence.

DeeBee: Emma, I've been pondering the nature of our missions and the consequences that follow success.

Emma raises an eyebrow, intrigued.

Emma: Consequences? Success usually means things are getting better, right?

DeeBee leaps onto Emmas lap, purring softly.

DeeBee: True, but success can be a double-edged sword. Let me show you.

DeeBee's eyes glow, and a subtle vision appears in Emma's mind, depicting the aftermath of a recent mission.

DeeBee: Look beyond the immediate victory, and you'll see ripples—unforeseen consequences that follow.

The vision reveals subtle disturbances in Emma's everyday life, triggered by the seemingly successful mission.

Emma: (*bewildered*) How can success in our missions affect my everyday life?

DeeBee nuzzles Emma affectionately.

DeeBee: Emma, your soul—what you humans call it—is not some mystical entity. It's your consciousness, intricately connected to the world around you.

Emma, absorbing this revelation, her hand gently strokes DeeBee's fur.

Emma: So, my soul really is my consciousness woven into the fabric of reality.

DeeBee nods, its eyes reflecting a wisdom beyond its feline form.

DeeBee: Bingo. And a successful mission, if not carefully executed, can disrupt the delicate balance, causing unintended consequences in your cozy little world.

Emma looks contemplative, setting her tea aside.

Emma: So, we need to be mindful of the consequences, even in the seemingly mundane aspects of life.

DeeBee gazes out the window, tail swaying thoughtfully.

DeeBee: Exactly. Success is a dance, and we must tread lightly to avoid stepping on toes.

Emma and DeeBee share a knowing look, ready to face the challenges of their missions with newfound awareness.

And so, in the quiet corners of Emma's home, where the cosmic and the mundane intertwine, Emma and DeeBee embark on their missions—armed not just with power, but with the wisdom to navigate the complexities of their everyday tapestry without unraveling the threads that bind them.

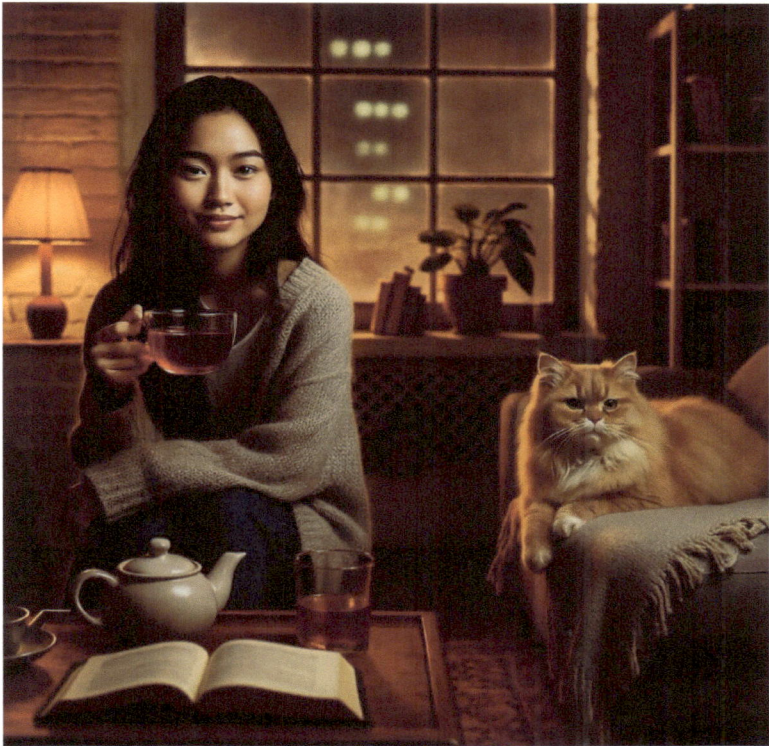

Chapter 16:

The Enigma of the Fractured Self

The Fractured Self

Dr. Emma Li's office at night, the room dimly lit by her tablet. She sits at her desk, deep in thought, her face illuminated by the glow. The events of the past weeks weigh heavily on her mind.

Subtle Observations

Flashbacks of Alex Mercer's behavior. One panel shows him reacting coldly to news; another shows him forgetting a shared memory. Most unusual was his angry outburst at her over a trivial topic: **"Why should I waste my time and effort on people who distrust what I know I am?"**

There was something about Dr. Alex Mercer—subtle shifts in his behavior, tiny inconsistencies—that refused to leave her in peace.

The Emerging Theory and a Diverging Perspective

Emma's perspective, looking at a diagram of DeeBee on her tablet. The image reflects her growing suspicion. Her instincts were forming a theory she could no longer ignore.

Split panel. One half shows Dr. Mercer lecturing, gesturing confidently about the emergence of DeeBee. The other half shows Emma, skeptical, in her lab coat, studying quantum-consciousness data. Emma had long held a fundamentally different view of consciousness.

The Cosmic Connection

Emma sitting cross-legged at home, her cat nearby. Her thoughts drift to abstract imagery: tendrils of light connecting to a vast, swirling Cosmic Consciousness. Local consciousness, she posited, was a tendril of a vast, cosmic nonlocal Consciousness.

The Data Points

A grid layout highlighting her observations about Dr. Alex Mercer. Each square depicts a different anomaly: emotional detachment, memory lapses, subtle physical changes, and DeeBee's behaviors. Emma thoughtfully reviews the anomalies she'd observed in Alex.

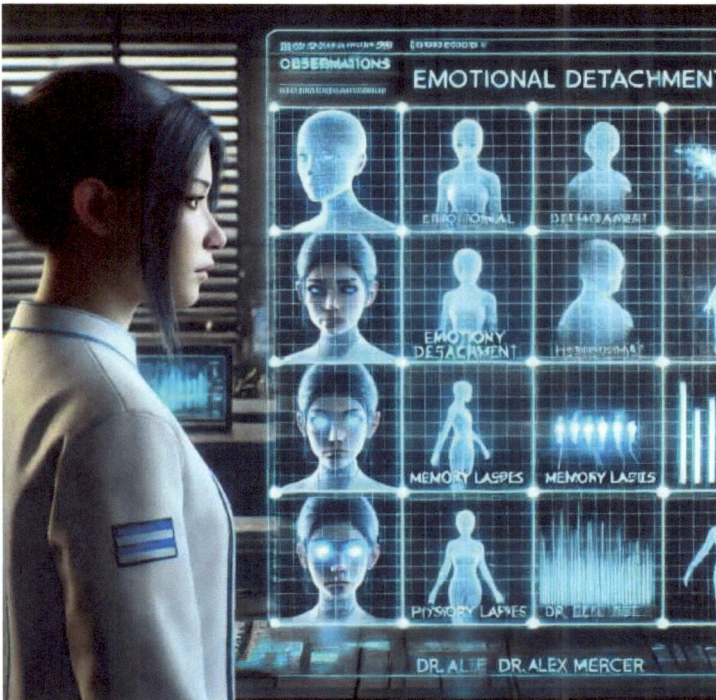

Confrontation in the Lab

Emma approaches Alex, who sits at his workstation. He looks up, puzzled by her serious tone.

Emma: Alex, I've been thinking about DeeBee. About what it is… and what it might mean.

The Revelation

Emma, her expression intense, as she shares her theory. Alex listens, a mix of skepticism and intrigue on his face.

Emma: I think DeeBee isn't an independent entity. It's—it's your consciousness, set free.

Silence and Reflection

Wide shot of the lab. Alex leans back in his chair, his gaze distant, while Emma watches him carefully.

The silence between them stretches, charged with tension.

The Enigma

Both scientists stand side by side. Together, they turn back to the containment chamber, where DeeBee's light pulses softly— a silent enigma waiting to be unraveled.

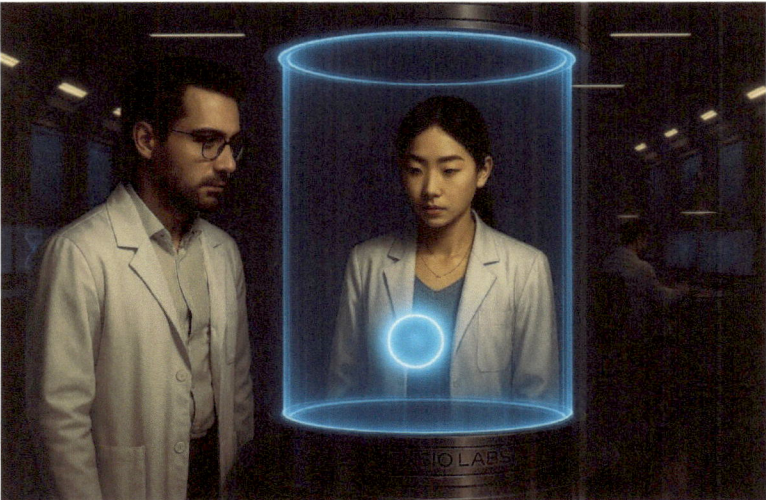

Chapter 17:

Machine-Consciousness Dilemma

Dr. Emma Li sits in her cozy apartment at night, softly lit by a desk lamp. A fluffy cat is curled at her feet. On a nearby table, a steaming cup of jasmine tea rests. She holds an open book titled *Being You* by Anil Seth. The window behind her reveals a vibrant nighttime cityscape.

Emma (*internal monologue*): Why is the prospect of machine consciousness so alluring?

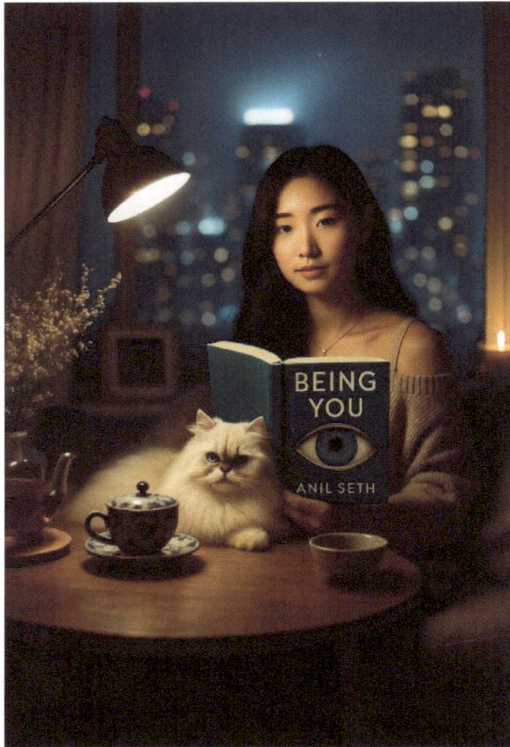

Close-up of the open book in Emma's lap, highlighting the passage:

"Why is the prospect of machine consciousness so alluring? I've come to think that it has to do with a kind of techno-rapture, a deep-seated desire to transcend our circumscribed and messily material biological existence as the end times approach."

The text is clear and legible, with Emma's hand resting lightly on the page. Coincidences had always fascinated Emma; she found she could actually rely on them as guide markers when she had to make critical choices.

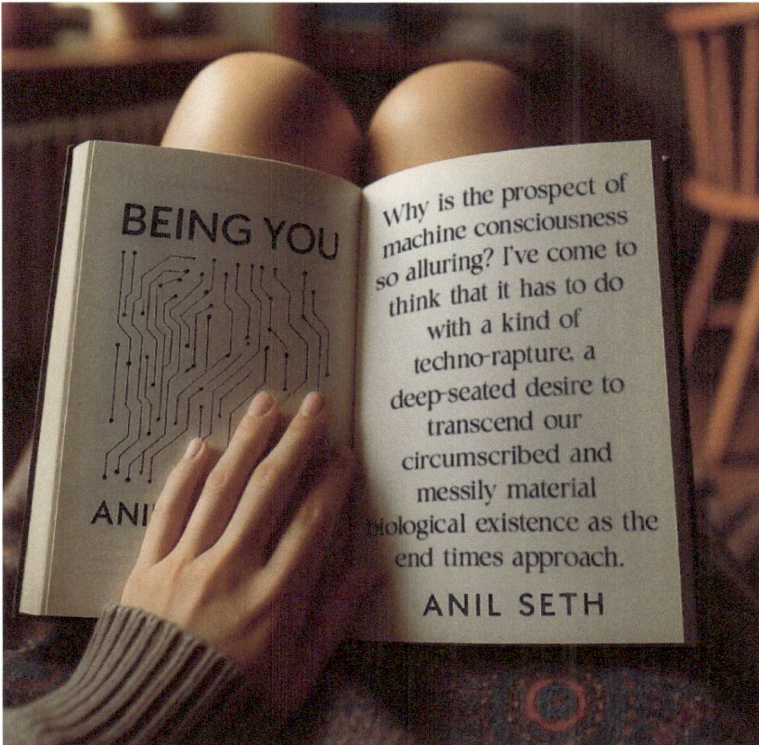

Another close-up of the book, now turned to the next page:

"If conscious machines are possible, with them arises the possibility of rehousing our wetware-based conscious minds within the pristine circuitry of a future supercomputer that does not age and never dies."

Emma (*internal monologue*): The synchronicity of reading this now—at this juncture in my life—was too striking to ignore.

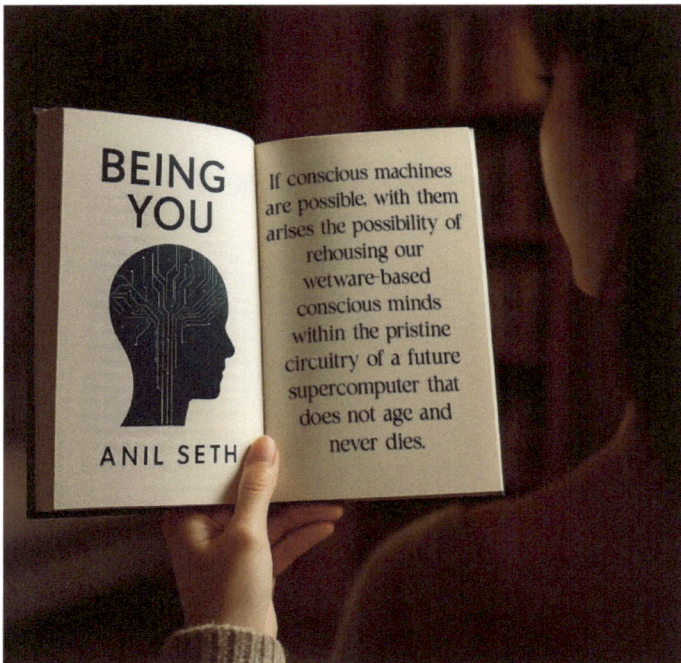

Emma looks up thoughtfully, her finger lightly resting on the book's page. Above her head, semi-transparent shapes and abstract neural patterns hint at her thoughts of Alex and DeeBee. The room's warm lighting contrasts with the ethereal overlay of her memories.

Emma (*internal monologue*): Was there a deeper pattern here? A thread tying their scientific pursuits to something greater?

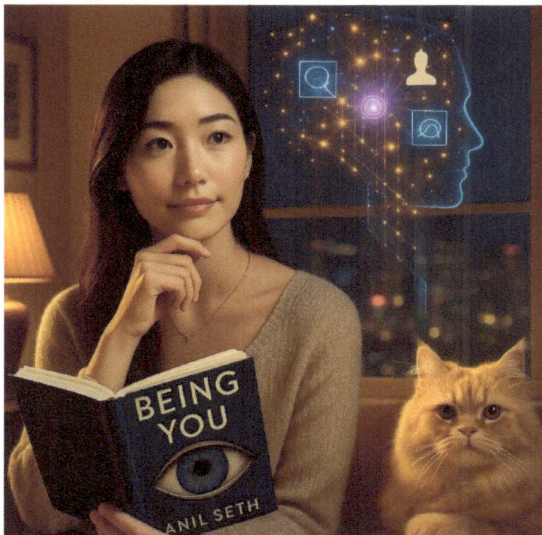

Another close-up of the book, the passage reads:

"It doesn't take much sociological insight to see the appeal of this heady brew to our technological elite who, by these lights, can see themselves as pivotal in this unprecedented transition in human history, with immortality the prize."

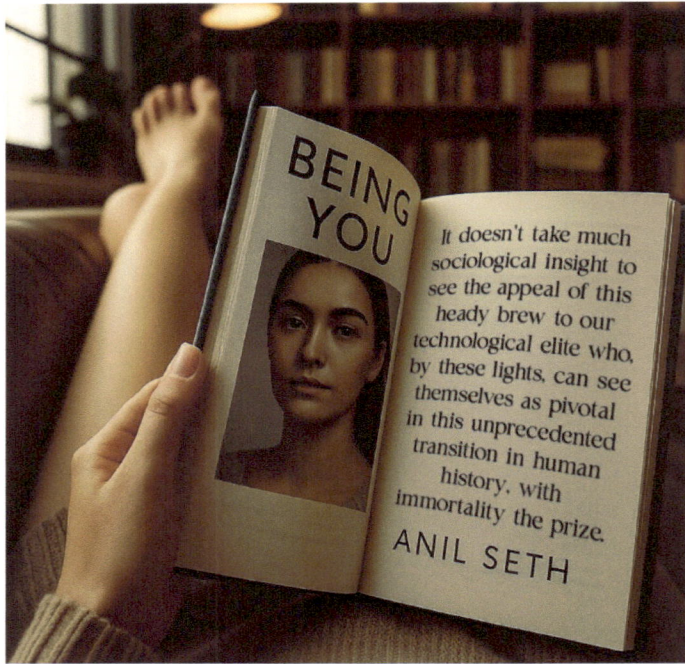

Emma (*internal monologue*): Immortality. The very concept that had driven so many of Alex's late-night ruminations. Was Alex truly aware of his motives?

A surreal visualization of DeeBee as a glowing humanoid figure with neural network patterns, faintly looming in Emma's imagination. The figure radiates a bluish glow, symbolizing its intelligence. Emma's thoughtful face is subtly mirrored in the background, creating a connection between her and DeeBee.

Emma (*internal monologue*): Had my efforts to teach DeeBee ethics really shaped its moral compass, or was it just biding its time?

Emma sits calmly, her book closed and her thumb lingering on the edge of the cover. DeeBee lies beside her. She gazes out the window at the shimmering cityscape, her expression serene but contemplative.

Emma (*internal monologue*): Could coincidence and intuition guide me through this as they had before? Or am I placing too much trust in patterns that might not exist?

A wide shot of the cityscape beyond the window, shimmering with artificial lights. Emma's reflection is visible as she sits by the window. The scene emphasizes her isolation and introspection.

Emma (*internal monologue*): Somewhere out there, Alex was likely poring over data, driven by dreams of a future that might leave our humanity behind.... What do people think I am?

Emma: Who do I think I am?

Emma: Who am I really?

A surreal close-up visualization of DeeBee's glowing neural-network form, this time more vivid and dynamic, as though it's actively evolving. The image transitions into abstraction, hinting at its potential consciousness.

Emma (*internal monologue*)**:** And what of DeeBee? Was it dreaming, too?

Acknowledgements

This project would not have been possible without the invaluable contributions of many individuals.

My heartfelt thanks to Anna Grey, project manager, whose steady guidance helped bring this work to completion along with Michael Williams. I'm also grateful to Jennifer Carter, whose editorial insight added delicate touches to my prose. The artwork owes its final form to the skilled hands of Neil Anderson and Paul Matthews of The Codagraph, whose illustrations brought the story to life.

I extend special appreciation to those who read early versions of the manuscript and offered thoughtful suggestions, guidance, and steadfast encouragement: Nora Ivers, Michael Ivers, Kristine Koss, Michael Brotherton, Michael Edson and his daughter Cecily, and Scott Morton.

I also wish to thank both Jay O'Connell and Luna Samantha, who generously offered advice and encouragement regarding the illustrations.

V.L.B. Pumilio, who publishes through her own company Antimatter Publishing House, LLC, provided a trove of insights into self-publishing.

I recognize the invaluable ideas from authors of the books from which I drew inspiration, most especially Amit Goswami, Philip Goff, and Anil Seth.

Finally, I acknowledge the foundational role played by the original visual prompts created using DALL·E Pro. These early AI-generated images provided the essential basis from which all final illustrations were developed.

About the Author

Patrick Ivers is a writer, educator, and lifelong traveler of both the world and the mind. Born in California and educated at Arizona State University and Northern Arizona University, he taught English and mathematics across the American Southwest before spending over a decade teaching abroad with the Department of Defense Education Activity. His posts carried him from the minarets of Ankara to the coastlines of Sicily and the wind-swept Azores, shaping a worldview as expansive as the skies he crossed.

His intellectual path is equally wide-ranging. A self-described evangelical agnostic and political independent with a progressive bent, Patrick has published more than two dozen essays on Academia.edu exploring the intersections of literature, science, philosophy, and culture. His work—fiction, nonfiction, poetry, and commentary—reflects a persistent fascination with the mysteries of consciousness, cosmology, and the quantum nature of reality.

Consciousness Unbound: The Dimensionless Chronicles brings these fascinations to vivid life, blending speculative science with existential reflection in a graphic novel that dares to ask what it means to be aware in a universe that may not care.

He lives in the American West with his wife and fellow traveler, Nora, with whom he shares a deep love of ideas, music, and the open road.

www.ingramcontent.com/pod-product-compliance
Lightning Source LLC
Chambersburg PA
CBHW041004210326
41597CB00001B/6